FORD AT WAR

LKW · Maultiere · PKW

Reinhard Frank

Advertisement for the Ford 3-ton truck, model V 3000 S (Type G 198 TS), in Luftwaffe service.

SCHIFFER MILITARY HISTORY
Atglen, PA

Advertisement for the "Ford of the homefront" — Type G 188 TG with wood-gas generator.

Sources
Specifications
Manufacturer's data
Factory publications
Trade publications
Archive material
Original photos by combatants

My special thanks go, besides to Gertrud, Stefan and Daniel, to Mr. Hans Thudt, without whose great help this volume could not have been written. Valuable photos and documents were kindly made available by: Siegfried Ehrt, Henry Hoppe, Randolf Kugler, N.CSSR, Alfred Otte, Ulrich Rohrbach, Helmuth Spaeter, Peter Taghon and others.

The drawing on the cover shows a Ford G 917 T of the 21st Panzer Division near Sidi Azeiz, North Africa, in November 1941.

Back cover: Ford 3-ton truck, model V 3000 S (Type G 198 TS), with infantry and anti-aircraft machine guns.

Translated from the German by Edward Force.

Printed in the United States of America.
ISBN: 0-88740-480-4

This title was originally published under the title, *Ford im Kriege*, by Podzun-Pallas Verlag, Friedberg.

Published by Schiffer Publishing, Ltd.
77 Lower Valley Road
Atglen, PA 19310
Please write for a free catalog.
This book may be purchased from the publisher.
Please include $2.95 postage.
Try your bookstore first.

We are interested in hearing from authors with book ideas on related subjects.

FOREWORD

The documentation at hand concerns the vehicles produced by Ford of Germany that saw service during World War II. Several Fords made outside Germany that also saw German service will also be noted.

With energy and a wealth of ideas, Henry Ford succeeded in creating a worldwide automobile empire. In the 1930s there were production or assembly facilities not only in Germany, but also in the USA (Dearborn MI), Britain (Dagenham), Canada (Windsor), Australia (Melbourne), South Africa (Port Elizabeth), New Zealand (Lower Hutt), France (Poissy and Asnieres), Holland (Amsterdam), Belgium (Antwerp), Rumania (Bucharest) and elsewhere.

In Russia, the newly founded firm of GAZ in Gorky began in 1932, with American help, to produce the "Russky-Ford."

The variety of models made worldwide was deliberately kept small, so as to be able to offer high-quality but reasonably priced vehicles. The vehicles built in the various countries were very similar in appearance; making a definite identification is often difficult.

A short Ford chronicle:
Henry Ford, born July 30, 1863, the son of Irish immigrants, built the first car of his own design in 1896 and founded the Ford Motor Co. in Detroit in 1903. As of 1903, the Model T was produced on the assembly line, and by 1927 fifteen million of them had been delivered, forming the majority of Ford production. Production of the Model A began in 1927, and in 1932 the first V8 was built, of which 6000 cars per day were produced in 1938. The twenty-five millionth Ford, and the four millionth V8, were built in 1937.

In Germany, the "Ford Motor Company A.G." was founded in Berlin in 1925 to import American-made Fords. In 1926 an assembly plant was built in Berlin-Plötzensee. This assembly plant was closed in 1931, when the Ford plant in Cologne-Niehl was opened. The technical facilities of the Berlin plant were apparently sold to GAZ (Gorki, Soviet Union). As of 1933, all four-cylinder Fords were recognized as being "made in Germany", since they were made extensively of German materials.

In 1935 the process of turning an assembly plant into an independent German automobile factory was completed. Until the beginning of World War II, the reliability of Ford vehicles was proved again and again in numerous long-distance trips and expeditions.

The firm's name was changed to "Ford Werke A.G." in July 1939. The yearly production for 1939 was 35,364 vehicles (including 18,219 passenger cars), made by 3847 employees. In 1940, 17,557 vehicles (4834 cars) by 3871 employees. In 1941, 16,243 vehicles (927 cars) by 3476 employees. In 1942, 14,992 vehicles (41 cars) by 4650 employees. In 1943, 17,472 trucks by 5711 employees. In 1944, 13,015 trucks by 5134 employees. In 1945, 2846 vehicles by 2745 employees. Truck production was resumed on May 8, 1945, the day Germany surrendered, a fact which appears to sustain assertions that the Ford works had been "spared" in bomb attacks.

For war service, the three-ton truck, "Maultier" (Mule) halftrack and "uniform heavy passenger vehicle" were built. All other types were included as "auxiliary vehicles", and all types of Ford cars and trucks were very popular among the troops on account of their reliability.

The model designations for Ford trucks can be understood on the basis of the model G 198 TS: G = Germany, 1 = model year 1941 (9 = 1939, 0 = 1940, etc.), 9 = 3.9 liter engine (1 = 3.6 liter, 8 = 3.2 liter), 8 = model, T = truck, S = standard 4x2 (G = wood-gas generator).

Special thanks are due to Mr. Hans Thudt of Oberursel, without whose documentary and photographic material this book could not have been prepared. Herr Thudt comes from Hermannstadt in Siebenbürgen (now Sibiu, Rumania), and has been collecting, since he was twelve years old, absolutely everything that dealt with Ford vehicles. In more than fifty years, he has built up the largest private collection on Ford models of the 1930s and 1940s, and has collaborated in numerous publications on civilian Fords.

TRUCKS

FORD MODEL "BB" TRUCK

The German Ford BB (four-cylinder 13/50 HP truck) corresponded fully to the American Ford BB ("proved worldwide") and was produced from 1932 to 1939. There were chassis for a 1.75-ton payload (with normal rear body) and a wheelbase of 3.34 meters, designated "BB-Light" (or Special), as well as for a 2.5-ton payload, with a wheelbase of 3.99 meters (BB-Long) and 3.34 meters (BB-Short). Both versions had the usual Ford dual rear wheels and differed only in the location of the spare wheel, which was placed in the middle under the rear body of the 1.75-ton but under the frame end of the 2.5-ton type. There were bodies "for every use": open trucks, buses, box vans. Beginning with the 1937 model year, minor bodywork modifications were made, particularly the introduction of straight headlights. Only limited quantities of these trucks were requisitioned for service with the troops.

TECHNICAL DATA (2.5-ton payload)

Engine	Ford BB
Cylinders	4 in line
Bore	98.4 mm
Stroke	108 mm
Displacement	3285 cc
Compression ratio	4.6:1, 7:1 for wood gas
Maximum power	52 HP at 2800 rpm
Carburetor	Solex vertical
Speeds	4 forward, 1 reverse
Top speed	75 kph
Turning circle	14 meters
Springs, front	1 transverse semi-elliptic
rear	2 longitudinal semi-elliptic
Four-wheel brakes	Mechanical
Wheel size	5"/20 L
Tires, front	6.00 transport -20
rear	7.00 transport -20X2
Track, front/rear	1422 mm/1642 mm
Wheelbase	3990 mm, 3340 mm
Ground clearance	210 mm
Fuel consumption	24 liters/100 km
Oil consumption	0.15 liter/100 km
Timing	Automatic

France 1940: Ford BB, 1937 model (WH-326315) with infantrymen on board passing a B 1 tank.

Right: A Ford BB, 1932 model (curved headlight mount) loaded on a railroad car — a requisitioned vehicle from Hannover province, used by the 86th ID (probably Artillery Regiment 186).

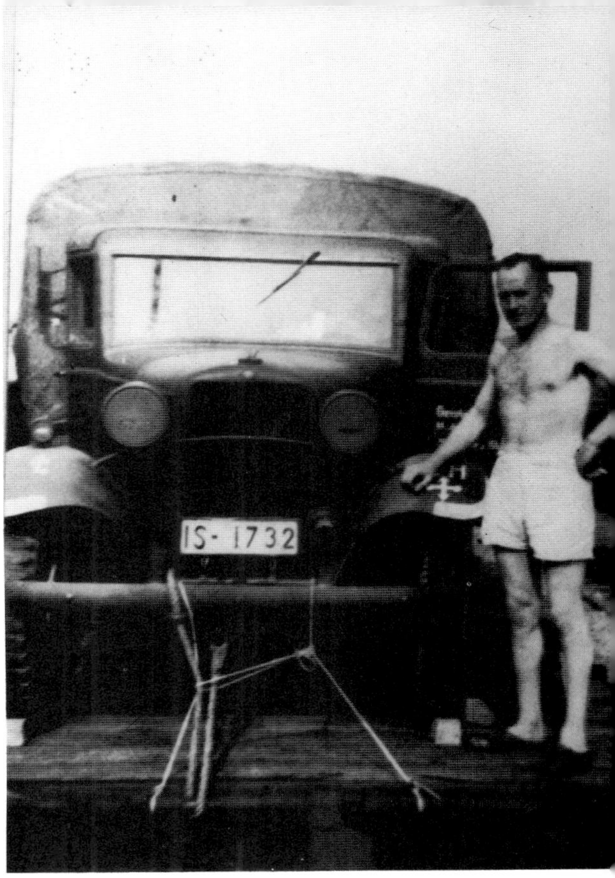

Below: A Munich Ford BB, 1937 model, during NSKK maneuvers. In the background is a Ford V8-51.

Below: A Ford BB, 1937 model, of I./JG 53, crashed with a broken axle, in May 1940.

FORD MODEL V8-51 TRUCK

In 1932 Ford (USA) put its legendary V8 engine on the world market. Both cars and trucks were powered by this powerful engine. In Germany the first V8 truck was a three-ton type called the V8-51 and built from 1937 to 1951. The chassis had a wheelbase of 3.34 (short) or 3.99 (long) meters. In the armed forces, and especially the NSKK, this model could be seen very often. There were civilian versions of every type one could want, while the version with open cab and rear body was intended for military use.

TECHNICAL DATA

Engine	Ford V8
Cylinders	8, in V form
Bore	77.5 mm
Stroke	95 mm
Displacement	3583 cc
Compression ratio	6.32:1
Maximum power	90 HP at 3800 rpm
Carburetor	Solex downdraft
Speeds	4 forward, 1 reverse
Top speed	83 kph
Turning circle	18 meters
Springs: front	1 transverse semi-elliptic
rear	2 longitudinal semi-elliptic
4-wheel brakes	Mechanical
Wheels: front/rear	5"-20 L/6"-20
Tires: front	6.50 transport-20
rear	7.25-20 extra x2
Track: front/rear	1422 mm, 1642 mm
Wheelbase	3990 mm, 3340 mm
Ground clearance	220 mm at rear axle
Fuel consumption	26 liters/100 km
Oil consumption	0.15 liter/100 km
Timing	Automatic

From "Comrade Motor Vehicle": A Ford V8-51 with single rear wheels and Belgian body, probably captured.

Left: Front end of the Ford V8-51, depicted in Ford's "Customer Service News."

Right: The Ford factory maintained close contact with employees who were drafted into the army. This photo, from the Ford factory newspaper of November 1940, shows employee Helmuth Ammon (x), who wrote of his experiences between Narvik and Tronso." Unfortunately, he wrote nothing about his unit, the load on the special trailer (searchlight, RRH, command equipment?), and nothing about the Ford V8-51 with open cab and unusually long chassis.

Left: The fuel column of an unfortunately unidentified unit stops with technical problems. A tripod-mounted MG 15 provides picturesque anti-aircraft protection. Note the different cabs on these Ford V8-51 trucks.

A Ford V8-51 belonging by its markings to Supply Column Unit 572 of the *Afrika-Korps* (emblem reversed). The shape of the cab, with rain gutter and stamped doors, is unusual, as is the low-mounted rear body with a wheel well.

Maikop, Russia, summer 1942: Open-cab Ford V8-51 trucks of 5./Railway Engineer Regiment 3.

FORD 3-TON MODEL 1939 TRUCK (Uniform Vehicle)

In 1938 Ford of America introduced the Model 817 T (and others) with a new oval radiator grille. This was used, slightly changed, by Ford of Cologne for its 1939-1941 3-ton trucks.

Four types were made; their engine numbers do not tell how many were actually made:

1. G 987 T, numbered from 5 370 801 to 5 373 000, 5 398 001 to 5 408 000, built 1939-41, 4 cylinders, 3.2 liters.
2. G 087 TG: numbered from 5 370 801 to 5 373 000, 5 398 001 to 5 408 000, built 1939-41, 4 cylinders, 3.2 liters, with wood-gas generators.
3. G 917 T, numbered from 5 129 853 to 5 135 000, 5 695 001 to 5 704 900, 6 190 001 to 6 197 927, built 1939-41, V-8, 3.6 liters.
4. G 997 T, numbered from 415 001 to 424 000, 528 501 to 530 000, 533 001 to 533 200, 534 001 to 535 000, built 1941, V-8, 3.9 liters.

The types built especially for the Wehrmacht were, according to D 669/15, given the designations of G 917 T St IIIa (3.6-l-Motor) and G 997 T St IIIb (3.9-l-Motor) and could not be told apart externally. In all, some 25,000 of them were built. In addition, 1500 Ford G 917 T trucks were built for Hungary, and some were fitted with Marmon four-wheel drive.

TECHNICAL DATA(open rear body)

	G 917 TSt IIIa	G 997 TSt IIIb
Engine	Ford V-8 3.6 l	Ford V-8 3.9 l
Power (factory data)	90 HP	95 HP
Bore	77.8 mm	81.0 mm
Stroke	95 mm	95 mm
Displacement	3613 cc	3917 cc
Compression ratio	6.16:1	5.9:1

Speeds	4 forward, 1 reverse
Springs: front	transverse semi-elliptic
rear	longitudinal semi-elliptic + secondary
Foot brakes	ATE hydraulic on all 4 wheels
Hand brakes	Mechanical cable, on rear wheels
Wheel size	6x20
Tire size	7.25X20 extra
Wheelbase	3988 mm
Front track	1475 mm
Length	6220 mm
Height over cab	2200 mm
Width	2240 mm
Turning circle	19 meters
Allowable load	3000 kg
Allowable gross wt.	5700 kg
Least sustained speed	6 kph in 1st gear
Top speed, 3550 rpm	85 kph
Fuel consumption	25.9 liters norm
Fuel capacity	105 liters

Left: Front end of the Ford G 917 T with rounded fenders, depicted in the "Customer Service News." Only the German version had the one-piece windshield shown here.

Above: Front view of a Ford G 917 T (G 997 T) of VIII Fliegerkorps (Commanding General Baron von Richthofen). By its infantry assault symbol, it is the vehicle of a fighter unit.

Right: The Ford G 917 T (G 997 T) of a Luftwaffe unit (the V may mean Versorgung: supply). The rabbit emblem indicates a JG 77 airfield. The truck was painted blue-gray.

Below: A Ford G 917 T (G 997 T) of the 21st Panzer Division in Libya, November 1941.

Left: After moving into the Baltic states, Stalin also occupied Bessarabia and northern Bukovina, parts of Rumania, in the summer of 1940. Antonescu wanted to reorganize his troops after the German system, and summoned German instructional troops (some 20,000 men under General der Kavallerie Erik Hansen) to Rumania in September of 1940. Here their Ford 3-ton trucks roll happily into Hermannstadt.

Right: Bulgaria, March 1941: A Ford G 917 T (G 997 T) of the Leibstandarte Division with number SS49815 and the tactical emblem of a small vehicle column (4th platoon?). Note the new-type flat fenders and the covered headlight mounts.

Left: Ford G 917 T (G 997 T) with flat fenders, probably vehicles of the 5th Light Division, Panzer Regiment 5.

Left: A Ford G 917 T (G 997 T) with open cab and curved fenders, part of a truck unit.

Below: This picture, unfortunately not very sharp, shows a fuel depot of the *Afrika-Korps* with a Ford G 917 T (G 997 T) with sheet-metal cab at left, and the same type with open cab at right.

Left: The Ford G 917 T (G 997 T) with open cab also existed with new-type flat fenders. The division emblem could not be identified.

Right: Dünaburg area, 1941: Ford G 917 T (G 997 T) trucks with open cab and flat fenders, of the 3rd SS Panzer Division "Totenkopf."

Left: Wheel trouble for a 3-ton Ford, WH-253190, of an unknown unit in Russia. Faulty photos such as this one are unfortunately common among private pictures but should not detract from their historical value.

Right: Ford G 917 T (G 997 T) Luftwaffe trucks. The first two trucks have an unusual type of cab with three door hinges.

Left: Medical corps ambulances (Kfz.31) on Ford 3-ton chassis (with single rear wheels. Made by the firm of Christian Miesen, only a few were built.

Right: The Ford 3-ton chassis was also fitted with a bus body. This one was used as makeshift housing after the war.

Above: The serious fuel shortage compelled vehicles back home to be converted to run on "native" fuels. All Ford trucks could be equipped with wood-gas generators. On G 917 T (3.6 l) and G 997 (3.9 l) eight-cylinder types, the Imbert gas cooler was mounted ahead of the radiator shell, the ribs of which had to be cut off. On the G 987 T (3.2 l) four-cylinder types, the gas cooler was located under the engine hood. This picture shows a G 997 T with wood-gas system, Rhine Province registration with red permit marking, offset container under the bumper, and gas generator with a diameter of 550 mm in back.

Left: Either a Ford G 087 TG with standard generator system or an updated G 987 T.

FORD THREE-TON TRUCK MODEL V 3000 S

Ford of America brought out the new-style type 018/098 T in 1940. Since Germany was not yet at war with the USA at that time and communication with the German Ford works was still going on smoothly, Cologne changed to the new type in 1941 — with slight changes.

The three-ton trucks with the 3.9 liter V-8 engines were called "V 3000 S" by the Wehrmacht. The four-cylinder types were intended for the homefront. Some 25,000 of them were built, plus unknown quantities of vehicles and spare parts in occupied France, Holland and Belgium.

The V 3000 S was certainly practical and reliable, but as a typical 4x2 civilian truck it left something to be desired in terms of off-road capability. Several British Commonwealth countries equipped their three-ton Ford trucks with front-drive units made by Marmon-Herrington to make them four-wheel-drive vehicles; there were even three-axle 6x4 and 6x6 types. A three-ton Ford with all-wheel drive was presumably also built in Germany, in a forward control version, as "G 198 TWA, V 3000 A, 1943-44." Unfortunately, this cannot be substantiated by factory documents.

The following models were built:
1. G 188 T, engine numbers from 5 409 151 to 5 423 800, built 1942-43, 4 cylinders, 3.2 liters
2. G 188TG, no. 5 409 151 to 5 423 800, built 1942-43, 4 cylinders, 3.2 liters, wood-gas generator
3. G 198TS, no. 424 001 to 425 000, 515 001 to 528 500, 530 001 to 533 000, 533 501 to 534 000, 535 001 to 539 000, 425 001 to 429 001, built 1942-43, 8 cylinders, 3.9 liters.
4. G 388TS, no. 5 424 001 to 5 433 000, built 1943-46, 4 cylinders, 3.2 liters, high radiator
5. G 388TG, no. 5 424 001 to 5 433 000, built 1943-46, 4 cylinders, 3.2 liters, wood-gas generator, high radiator
6. G 398TS, no. 429 001 to 435 000, 539 001 to 560 190, 671 001 to 676 000, built 1943-46, 8 cylinders, 3.9 liters, high radiator

The models of the last war years generally had uniform cabs, wooden body sides, simplified (flat) fenders and the high radiator, with additional louvers in the engine hood.

February 1943, Kharkov area: a Ford V 3000 S field kitchen truck brings food to tank crews. Note the wide eastern tracks and the bulkhead armor of the Panzer III M. This picture is of the 1st Battery of the "Grossdeutschland" Army Flak Unit.

TECHNICAL DETAILS
(Open Truck)

Engine	**Ford V-8 3.9 liter**	**Wheel size**	**6x20**
Power at 3500 rpm	**95 HP**	**Tire size**	**190x20**
Bore	**80.95 mm**	**Wheelbase**	**4013 mm**
Stroke	**95-25 mm**	**Track: front**	**1650 mm**
Displacement	**3922 cc**	**rear**	**1652 mm**
Compression ratio	**5.9:1**	**Length**	**6390 mm**
Speeds	**5 forward, 1 reverse**	**Height over cab**	**2175 mm**
Springs: front	**longitudinal semi-elliptic**	**Width**	**2250 mm**
rear	**Longitudinal semi-elliptic + secondary**	**Turning circle**	**13.95 meters**
		Allowable load	**3310 kg**
Foot brakes	**ATE hydraulic on all 4 wheels**	**Allowable gross weight**	**5830 kg**
		Lowest sustained speed	**6 kph in 1st gear**
Hand brake	**Mechanical cable on rear wheels**	**Top speed at 3350 rpm**	**85 kph**
		Fuel consumption	**27 liters/100 km**
Wheels	**Disc with flatbed rims**	**Fuel capacity**	**110 liters**

1 Trailer hitch
2 Secondary spring
3 Spring support
4 Rear spring
5 Rear spring mount
6 Frame
7 Direction indicator
8 Side light
9 Cab
10 Trailer symbol
11 Windshield wiper
12 Rear-view mirror
13 Upper hood
14 Headlight
15 Hood fastener
16 Blackout light
17 Rear axle housing
18 Hand-brake cable
19 Half-shaft
20 Driveshaft
21 Fuel filler
22 Muffler
23 Running board
24 Fender
25 Bumper
26 Radio antenna
27 Number plate
28 Tow hook

Typical Russian road conditions: A Ford V 3000 S near Sublzov/Rshev in October 1942. A first-type truck with "model designation S" on the grille and Ford emblem "decoration right and left" on the lower edge of the engine hood. The helmet emblem of the I.D. (mot) "Grossdeutschland" and the tactical symbol "large vehicle column for fuel (50 cbm)" can be seen.

Left: A Ford V 3000 S of the 1st Battery, Army Flak Unit "Grossdeutschland" near Rshev in 1942.

Right: A motorcycle rifle battalion of the 2nd SS Panzer Division "Das Reich" in the Gshatsk area, September 1941. On the other side of the road, Ford 3-ton trucks with different cabs pass them. The first Ford has a divided windshield and was probably captured.

Left: These soldiers made their way from Czechoslovakia to the US Zone at the end of the war and are being cared for by a Red Cross nurse. In the background is a bus (probably an operations vehicle) on the 3-ton Ford chassis.

Above: Sicily, June 1943: Vehicles of the 4./Paratroop Panzer Artillery Regiment "Hermann Göring"; in front is a Ford V 3000 S, behind it a Ford G 917 T (G 997 T) with open cab.

Below: A Ford V 3000 S tanker semi refueling a Heinkel He 111 during the Battle of Britain. Unfortunately, no information on this interesting model is available. Another vehicle of this type is known to have been at an airfield in southern Russia in 1943.

Right: Factory photo of a Ford V 3000 S "new type" truck with flat fenders, central hood fastener, no hood ornament and one side antenna. The arrow at the lower right of the radiator grille indicates the Type 92 special winter equipment.

Left: A Ford V 3000 S "new type" truck destroyed during the Ardennes offensive. Its small headlights are typical of the last war years.

Right: The final version of the Ford V 3000 S with flat fenders, U-shaped bumpers, antenna on the right fender, central hood fastener and no "nose" of the hood.

Left: Copenhagen, 1944: A Ford V 3000 S with wood-gas generator, flat fenders and uniform cab. The gas generator and firewood are under a cover behind the cab.

Right: While German soldiers often had to walk because of fuel shortages in 1944, these Americans drove a captured Ford V 3000 S into Rheims! The vehicle has a shortened bumper and bears white star emblems. The single-axle trailer also appears to be of German origin.

Left: In 1943 the Ford V 3000 S appeared with a larger radiator, requiring additional louvers in the front of the engine hood and two ventilator flaps.

Above: Factory photo of the Ford V 3000 S (G 398 TS) with high radiator. Note the flat fenders, hood fastener and U-shaped bumper. In the background are trucks with the uniform cab.

Below: Truck production began again at Ford on the day Germany surrendered. This vehicle with high radiator and rounded fenders (!) has holes bored for blackout lights. Wooden wheels were used until delivery because of the shortage of tires.

Ford V 3000 S (G 398 TS) with high radiator, uniform cab, and a radiator guard which was not generally introduced. Note the antenna on one side and the towing hook and eye.

MAULTIER

HALFTRACK 2-TON TRUCK
Ford "Maultier" (Mule), Model V 3000/S/SSM

The bad road conditions in Russia, particularly during the mud season, were an insoluble problem to the 4x2 S types. One unit of the Waffen-SS (probably the 2nd SS Panzer Division) equipped a 3-ton truck with adapted Carden-Loyd tracks (captured from the British) in place of its rear wheels in the winter of 1941-42, thus improving its off-road capability considerably. This halftrack running gear was introduced officially and was the same for all so-called "Mules" built by Ford, Opel and KHD.

The Ford production version basically corresponded to the Type G 198/G 398. In the case of the D 666/10 there was also the possibility of shipping the necessary components to the troops and transforming available 3-ton trucks into powerful "Mules" (though they had only a 2-ton load limit). Several developments to replace the SSM (= SS Maultier) running gear were in progress, but they were never used for reasons of preserving uniformity. By the war's end, almost 15,000 Ford "Mules" (some 1000 of them in France) had been produced, most of them with open rear bodies as Sd.Kfz.3b.

The steering was very simple in form: For wide-radius curves, steering the front wheels was sufficient. For sharp curves and off the road, the driver had to "pull carefully" on one of the two hand-brake levers (which worked on the right or left track via cable), give some gas, and (original text) "move the steering wheel into the position required for the turn."

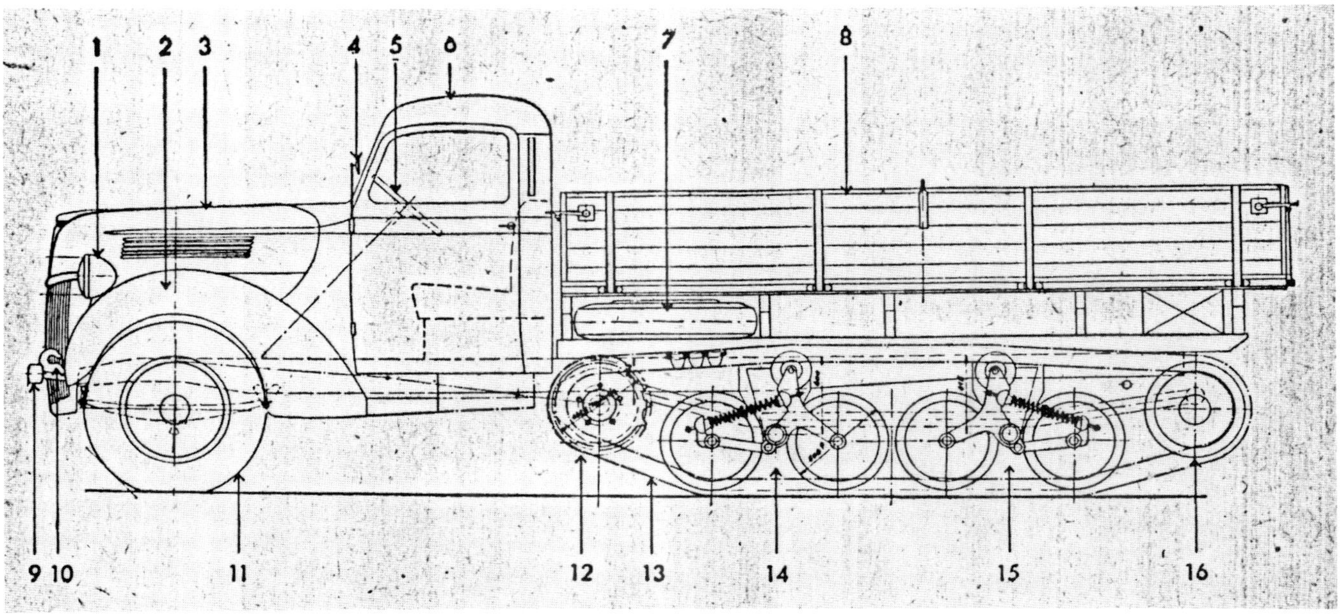

3-ton truck with tracks, Ford V 3000 S/SS M (Maultier)

1 Headlight
2 Fender
3 Upper hood
4 Rear-view mirror
5 Steering wheel
6 Cab
7 Spare wheel
8 Rear body
9 Bumper
10 Towing hook
11 Front wheel
12 Drive wheel 13 Track 14 Front road wheel 15 Rear road wheel 16 Leading wheel

TECHNICAL DATA
(SdKfz3b)

Engine	Ford V-8 3.9 liter
Bore	95.25 mm
Stroke	80.95 mm
Displacement	3924 cc
Compression ratio	5.9:1
Sustained power	83 hp at 3000 rpm
Maximum power	95 HP
Clutch	Ford single-plate with mechanical dampers, weights
Speeds	5 forward, 1 reverse
Speeds at 3000 rpm:	
1st gear (off-road)	4.8 kph
2nd gear	8.0 kph
3rd gear	12.8 kph
4th gear	23.5 kph
5th gear	39.6 kph
On paved road	25 kph
Low sustained speed	4 kph
Running gear:	
Attached to vehicle	Secondary frame
Drive wheels	2
Number of teeth	16
Leading wheels	2
Road wheels	8
Suspension	Swinging arms
Springs	Coil
Return rollers	4
Track tension	Leading wheels swing
Links	83
Lubrication	None
Foot brakes	ATE hydraulic on front wheels and tracks
Steering brakes	Mechanical cable on tracks
Activation	Two hand levers
Front wheel size	6"x20
Tire size	190-120 or 7.25x20e
Steering type	Worm gear, VDO position indicators
Front track	1650 mm
Rear track	1790 mm (from middle of track)
Overall length	6385 mm
Overall width	2250 mm
Overall height empty	2100 mm over cab
Turning circle	19 m -/15 m + steering brake
Allowable weight	5860 kg
Load limit	2000 kg
Trailer limit	2000 kg
Ground pressure:	
on soft ground	0.9 kg/sq. cm
on paved road	3.5 kg/sq. cm
Rolling pressure	591 kg
Wading ability	440 mm
Fuel consumption	50 liters/100 km
Off-road consumption	130 liters/100 km
Fuel capacity	110 liters

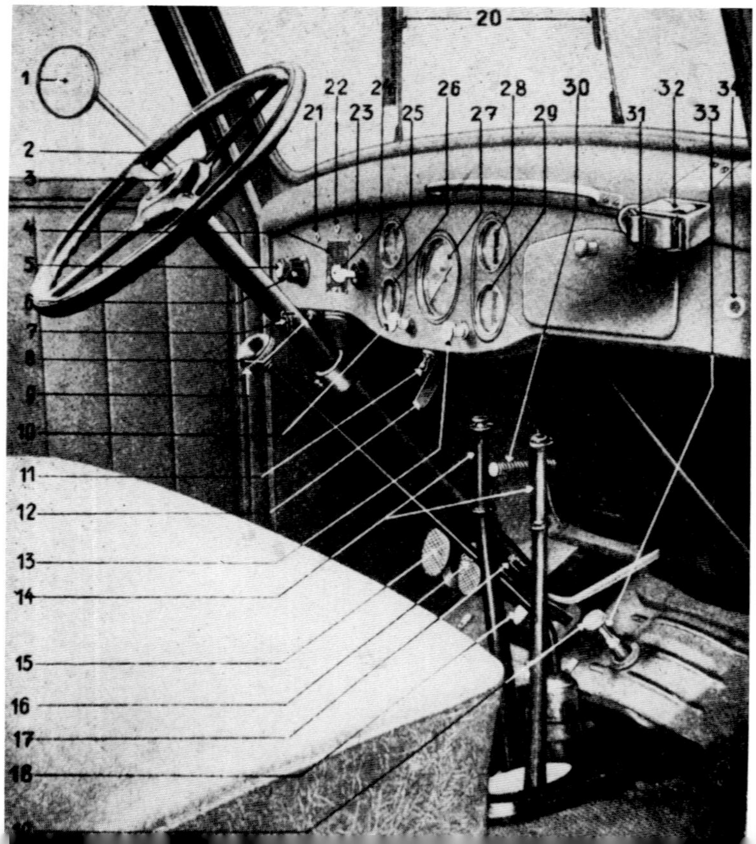

1 Rear-view mirror
2 Steering wheel
3 Horn button
4 Gear lever
5 Starter button
6 Night marching switch
7 Wiper switch
8 Dashboard light switch
9 Shift lever
10 Hand throttle
11 Directional light switch
12 Ventilator lever
13 Choke
14 Hand brake levers
15 Clutch pedal
16 Brake pedal
17 Dimmer switch
18 Gas pedal
19 Footrest
20 Windshield wiper
21 Load indicator
22 High beam indicator
23 Directional indicator
24 Ignition and light switch
25 Water temperature gauge
26 Oil pressure indicator
27 Speedometer/ odometer
28 Voltage indicator
29 Fuel gauge
30 Oil pump pedal
31 Storage box
32 Rifle holder
33 Gas pedal block
34 Plug

Above: Factory photo of a Ford "Maultier" WH-740663, with additional Rhine Province marking, road wheels with four round holes, leading wheels with openings and reinforcing ribs

Below: Ford V 3000 S/SSM (Maultier) WH-818675, leading wheels with openings but no ribs.

Right: Ford V 3000 S/SSM (Maultier) WH-740663 during testing.

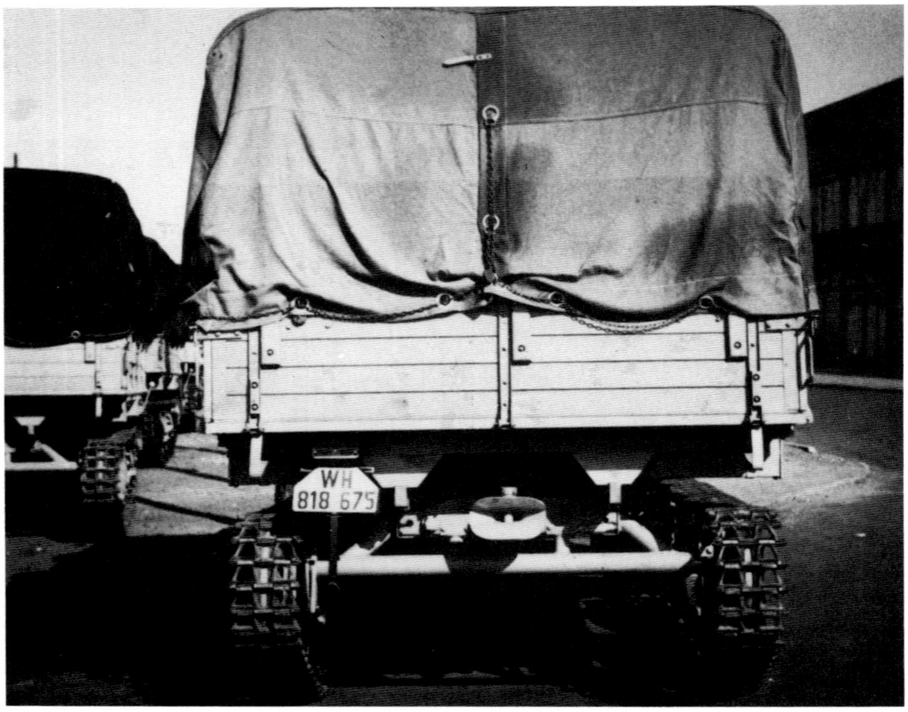

Left: Rear view of the Ford "Maultier." The auxiliary frame that holds the running gear is easy to see.

Below: Drawing of the Ford "Maultier" in its final form, with high radiator, flat fenders and uniform cab.

Left: Ford "Maultier" as a fuel delivery truck of the staff and supply company of Panzer Regiment 1 (1st Panzer Division) in the Bihar Udvari area, Hungary, October 1944.

Right: A Ford "Maultier" of the foglaying troops loaded with containers (2 x 900 liters) of fluid. These containers were made for the medium spray truck, Sd.Kfz.11/3, and fortunately never saw service. Note the spoked road wheels and the shape of the leading wheel.

Left: Ford "Maultier" (with spoked wheels) of the SS Pionier Battalion 9 in southern France in 1944, with an air compressor on hard rubber tires.

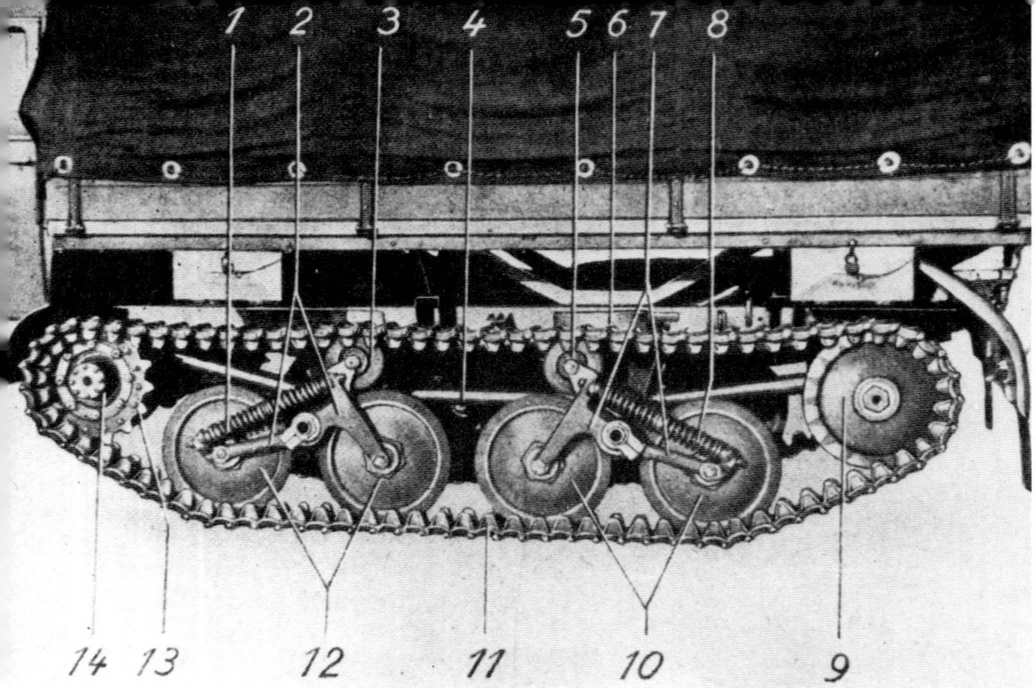

Left: running gear with track (D 666/409, July 23, 1943)

Front truck spring
2 Rear truck
3 Front return roller
4 Secondary frame
5 Rear return roller
6 Track bolt deflector
7 Rear truck
8 Rear truck spring
9 Leading wheel
10 Road wheels in rear truck
11 Track
12 Road wheels in front truck
13 Drive wheel teeth
14 Drive wheel hub

Right: A Ford "Maultier" (flat fenders), with disc road wheels and 20mm Flak 38 gun ahead of the shrapnel shield.

Left: Factory photo of a Ford V 3000 S/SSM, note the flat fenders with reinforcing ribs.

Right: An abandoned Ford "Maultier" is examined by British and French soldiers.

Left: A Ford "Maultier" with flat fenders in Budapest, late 1944.

Right: A destroyed Ford "Maultier" in the Ardennes offensive.

TRUCKS

THE HEAVY UNIFORM PERSONNEL CARRIER, FORD MODEL

From 1939 to 1941, Ford of Cologne built some 2000 "Uniform Chassis II for Heavy Passenger Vehicles." All of these vehicles had all-wheel drive and a Ford V-8 3.6-liter engine producing 78 HP at 3600 rpm. (Note: The performance of the 3.6-liter engine is always stated in government publications as 78 HP, also for other types; factory data list 90 HP at 3800 rpm.) Although it was supposed to be one uniform vehicle, there were nevertheless four different types:

— Type EGa: Support axle, four-wheel steering
— Type EGb: Support axle, front-wheel steering
— Type EGd: no support axle, four-wheel steering
— Type 40: No support axle, front-wheel steering

The heavy uniform vehicle had become quite heavy and ponderous, but its running gear turned out to be too weak and vulnerable. The Type EGb was to have been used particularly as a "carrier for various bodies"; thus special bodies for it will be shown below. It is impossible to tell it from Horch vehicles externally. On this subject, the reader is referred to the "VW AT WAR Vol.'s I and II" in the series, in which the technical data is also included.

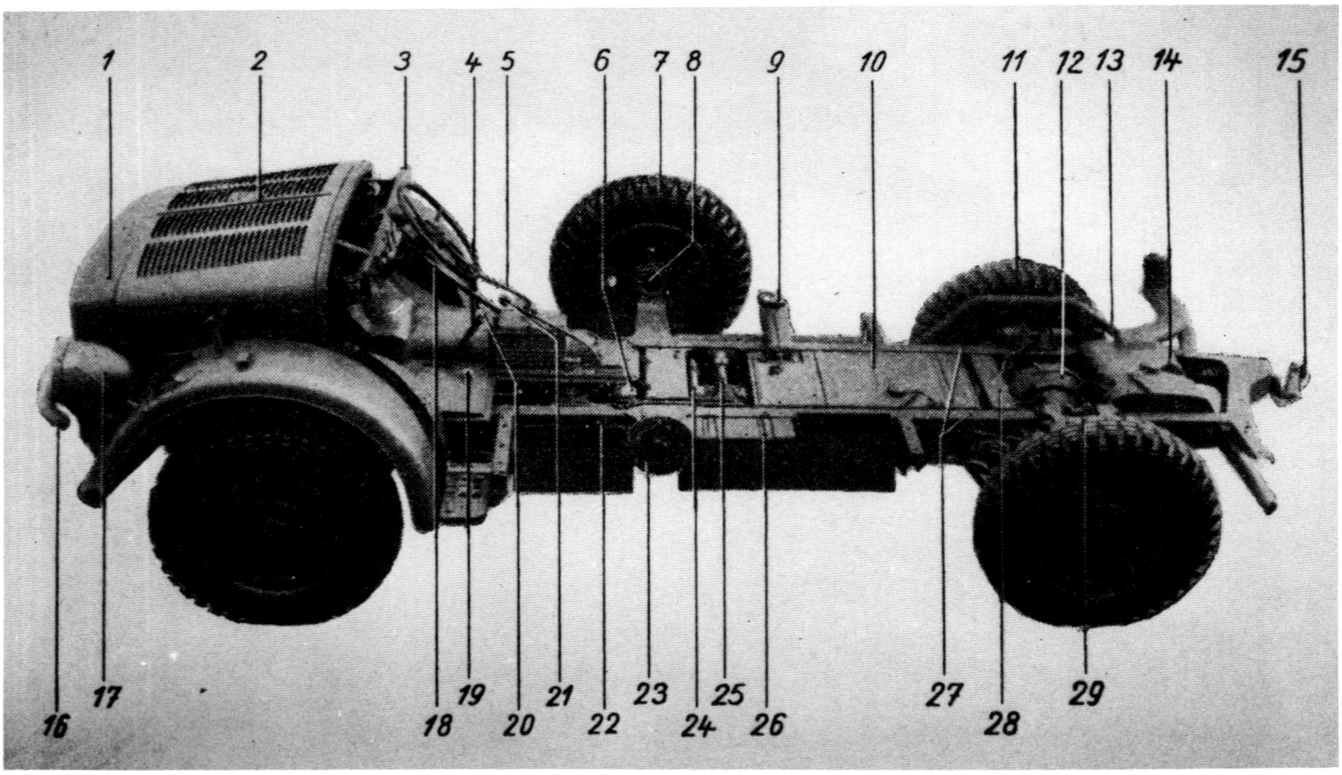

Uniform Chassis II for heavy passenger vehicles, Type EGa, from D 664/1:

1 Radiator shell
2 Engine hood
3 Dashboard
4 Steering wheel
5 Gearshift lever
6 Brake line
7 Spare wheel
8 Support axle
9 Fuel filler
10 Auxiliary fuel tank
11 Wheel
12 Rear differential
13 Exhaust pipe
14 Main fuel tank
15 Trailer hitch
16 Bumper
17 Width indicator
18 Hand brake lever
19 Fuel switch (main/aux. tank)
20 Steering switch lever
21 Equalizer cutoff switch
22 Toolbox
23 Spare wheel flange
24 Steering switch coupler
25 Steering connector
26 Battery case
27 Frame member
28 Cross member
29 Upper suspension member

Right: Some of the uniform chassis for heavy passenger vehicles were fitted with special bodies for 20mm Flak 30 guns, and with dropping grid sideboards. This vehicle belonged to the 7./ General Göring Regiment (absorbed into the Führer Flak Unit in 1942). Whether it used a Ford or Horch chassis cannot be determined, and no serial number is known.

Left: Self-propelled 20mm Flak gun on heavy uniform passenger vehicle chassis, of the General Göring Regiment. The gun crew wears aviation coveralls, as was customary in the 9./RGG.

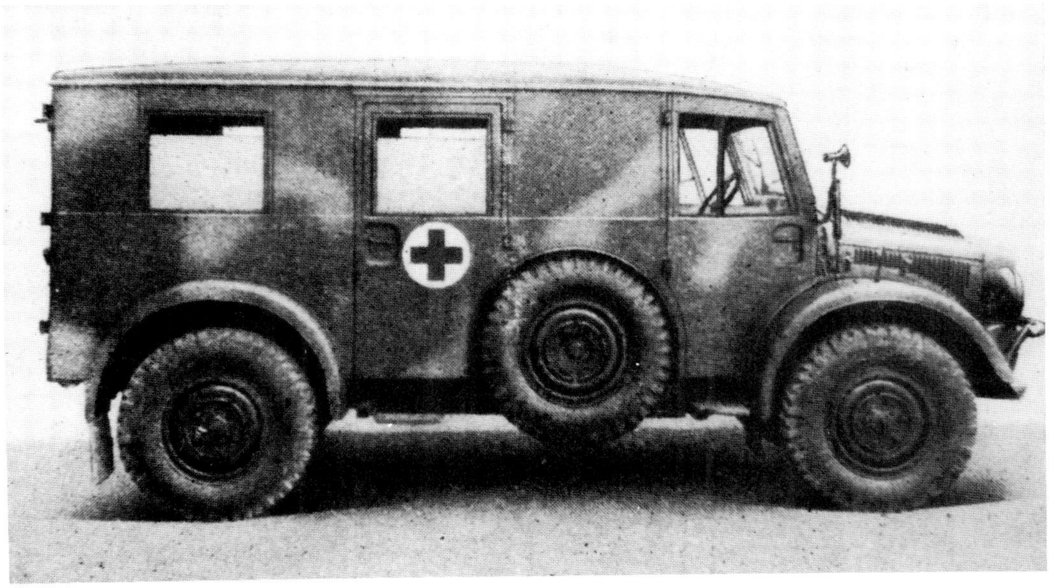

Right: Ambulance (Kfz.31) with uniform chassis II for heavy passenger vehicles.

Above: The first published picture of the "uniform chassis II for heavy passenger vehicles with cabriolet body", known for short as the commander's car. The manufacturer of the chassis and body has not yet been identified, but we wanted our readers to see this rarity. WL-483591 belonged to the Hermann Göring Panzer Division and bears the pennant of a Flak commander (?).

Left: The same commander's car, but with a general flagstaff, as introduced in February of 1941.

FORD CIVILIAN-TYPE PASSENGER CAR (o)

The civilian Ford cars had a good reputation for rugged quality and were, to the sorrow of their owners, requisitioned without exception for war service. Some of them were rebuilt (with makeshift box bodies) into radio vans, ambulances and the like. With welded-on equipment boxes, various types were turned into Kfz.15 (makeshift), though its off-road capability was very limited. The small "Eifel" was preferred on the homefront because of its rigid axle and turned into a towing tractor (capable of towing 5.5 tons at 20 kph) with the addition of an auxiliary towing axle. In the "Ford News" of the time there are personal experience reports that express high praise of Ford cars in service at the front, even under the worst road conditions.

Right: Another picture from "Camera Vehicle": A war correspondent stands on a Ford V8 Special two-door sedan. What would its owner have said? In the background is a Ford V8-51.

Below: A Ford Eifel cabrio-sedan, 1937 model (Stuttgart registration), of the Panzer Group Hermann Hoth, seen in the Smolensk in August 1941. On the other side of the road are two Panzerjäger "4.7 Pak (t) on Renault R 35 Tank (f)"; the triangle symbol could be that of the 4th Army.

Left: Ford V8-48 cabriolet, four windows, German body, 1937-39. with an as yet unidentified Luftwaffe emblem.

Right: Italy, 1944: Ford Eifel sport two-seater, 1938-39 model (roadster), of the Waffen-SS near an anti-aircraft gun position. The car bears camouflage paint and is, as usual, equipped with night marching equipment and a shovel.

Left: Ford Eifel sedan, built 1937-1939, of an unknown unit.

Above: Ford V8 luxury cabriolet, built 1938-39, of an intelligence unit, still with civilian registration for Berlin. The square on the right fender has not been identified.

Below: Ford V8 luxury cabriolet, built 1938-39, probably the same vehicle as above, seen from the left side. Note the typical rear wheel cover with the V8 symbol.

Left: A Ford V8 Special four-door sedan with Ambi-Budd body, built 1938-39, rebuilt into a radio car, Kfz.15 (makeshift). A frame antenna and the spare wheel are on the roof, and a box for equipment has been welded onto the fastback body.

Right: Likewise a Ford V8 Special, as above, but without antenna, probably the car of a propaganda company, registered WH-121581.

Left: A Baden Ford V8 four-door sedan with Ambi-Budd body, built 1937-38, of an intelligence unit. On the roof are attachments for a frame antenna and the spare wheel, and a box for equipment is on the rear end.

Right: Radio truck (Kfz.17) on Ford V8 chassis, with body by the Papler firm of Cologne.

Left: This Ford V8 sedan was turned into a Kfz.17 (makeshift) with a box van body by Papler of Cologne. The vehicle belonged to the staff of Panzerjäger Unit 38, 2nd Panzer Division.

Right: Ford V8 Special sedans turned into limbers to tow 20mm Flak 38 guns. These bodies were also built by Papler of Cologne.

Left: The owner of this four-door Ford V8 Special 1938-39 sedan with body by Ambi-Budd might have wept to see the cameraman with hobnailed boots standing atop his car at Dunkirk in 1940 taking films for the weekly newsreels. A sign of stability! Note the steps and handholds.

Below: Prototype of a Ford V8 Special with Imbert wood-gas generator for the homefront. The supply of wood was carried on the roof and could be reached by steps. The generator was covered very thoroughly; probably there was as yet no shortage of materials! The spare wheel, without a tire at the moment, was housed in a well on the left front fender. Series production did not take place.

CAPTURED FORD VEHICLES

A book could be written about the many captured Fords. The Wehrmacht, with its chronic lack of motor vehicles, naturally put to use all of the vehicles captured in France, Greece, Africa and elsewhere. All the better, many Fords corresponded fully to the German types and could be easily maintained. Some units, especially those of the NSKK, were equipped almost exclusively with captured vehicles and "got on well" with them. Besides the captured vehicles, naturally all the production facilities in France, Belgium and Holland were put into German service. The involved firms could supply no information as to the extent of vehicle and spare-parts production.

Right: A drawing from "Comrade Motor Vehicle", with an American Ford 1939 model in the background.

Below: In Holland, the firm of DAF equipped American Fords with open bodies and four-wheel drive to serve as towing tractors. Here is a Ford 91Y, modified by DAF, with the 1st Mountain (Gebirgs) Division in southern Russia in 1942.

Left: This four-wheel-drive Kübelwagen, based on the one-ton chassis of the 1939 US Ford, had a V8 engine (according to its hubcaps) and was exhaustively tested at Cologne. Unfortunately, no information on this interesting vehicle is available.

Right: A Ford of the *Afrika-Korps,* with the grille of the 1938 American model (817 T, etc.), wide front tires, position lights and open cab. The plaque on the grille may be that of the Amsterdam Ford assembly plant.

Left: It is not known why this Ford of JG 27 came to grief. It is probably a captured 1938 US model with its typical three vertical grille bars.

The great production rate of military vehicles in Canada is little known in Germany. By the end of the war, over 850,000 wheeled vehicles of all kind had been built there. The standardized "CMP" (Canadian Military Pattern) attained a state of perfection that was not equaled in the USA. The Canadian Ford was ugly, to be sure, but it drew high praise from the troops for its good quality. At the end of 1941, Army Engineer Battalion 659 (mot) in Athens had rebuilt fifteen Canadian Ford F15's into Kfz.15 (makeshift) according to plans and sent them on the march to Army Pionier Battalion 666 in Russia (recognizable here are WH-663581, 470618 and 537334). The emblem of Army Pionier Battalion 666, a circle of three 6 numerals, can be seen.

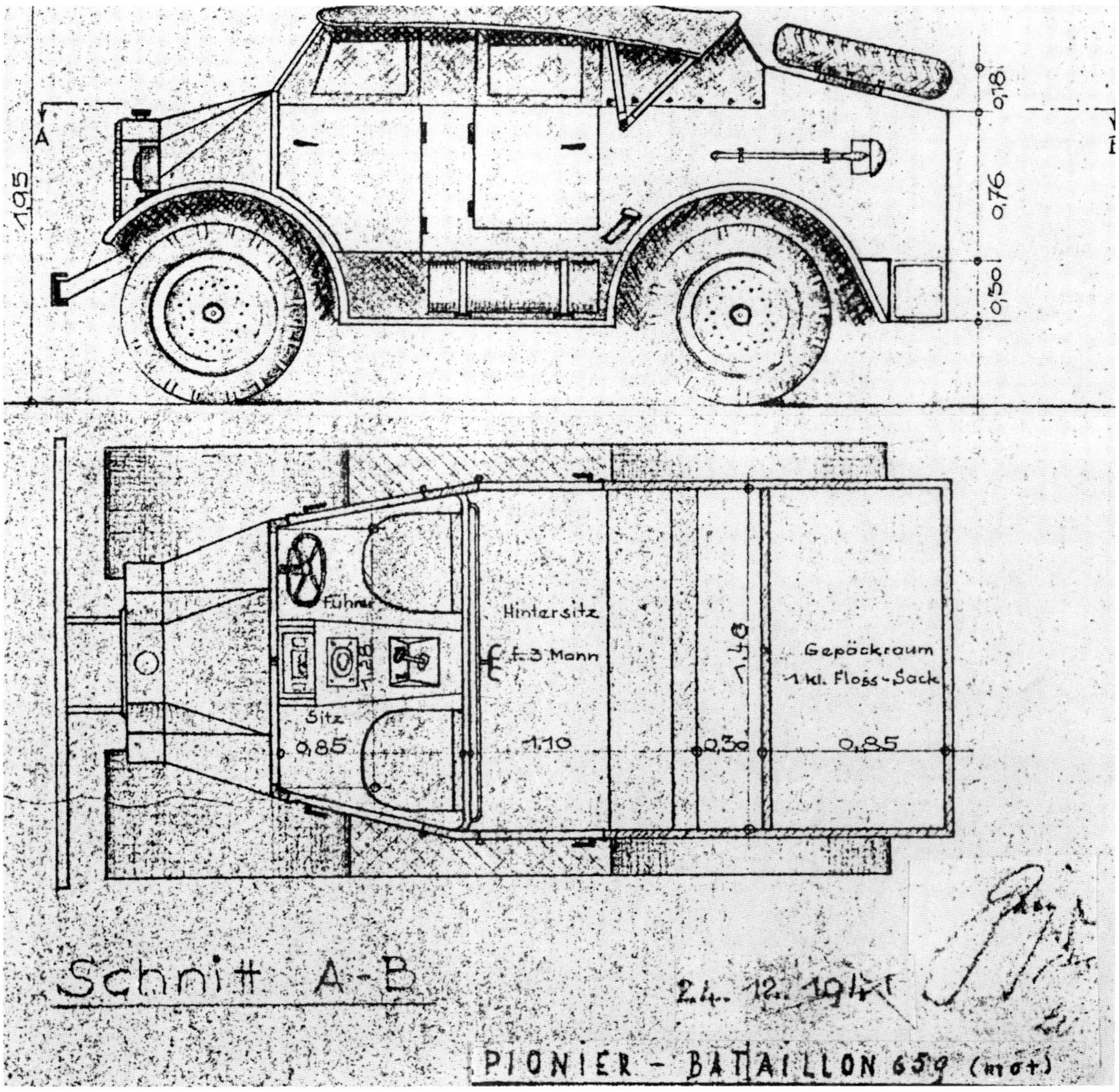

Above: This Canadian Ford F60 was used by the 13th Company of the "Brandenburg" Instructional Regiment for adventurous commando operations behind the enemy front. Note the German blackout lights.

Below: A Canadian Ford F60 in service with the *Afrika-Korps* in summer 1942. According to the unit marking "76", this vehicle formerly belonged to the British 12th Lancers, 1st Armoured Division.

Right: In 1934 Ford of France bought the Mathis firm in Poissy and operated as "Matford." In addition to cars, Matford produced 3.2 ton (01 W) and 5-ton (F 917 WS) trucks. The five-ton type was built in great numbers for the Wehrmacht from 1940 on, but their quality and reliability were much criticized by the troops. Here is a vehicle of the 3rd Infantry Division (mot) that broke down in Stalingrad. The tactical symbol indicates the 2./Pionier Battalion 3.

Left: In 1943 the grille design of the Matford F 917 WS was changed. Here Wehrmacht vehicles are ready for acceptance in Poissy.

Right: A Matford F 917 WS with wood-gas generator, with a Luftwaffe unit in France, 1944.

Left: The one-ton Ford 91 Y/DAF (4x4) Belgian Army towing tractor for the 47 mm cannon was equipped with light armor by the firm of Ragheno in Meghelen. This captured vehicle is in service with the I./IR 11; Battalion Commander Major Weiss is sitting beside the driver.

Right: In addition to trucks, a goodly number of cars were captured. This 1937 US Ford V8 Model 78 with whitewall tires (!) served with the II./JG 54 in France in 1940.

Left: Russia, October 1941, 5./Railroad Engineer Regiment 3, not a Ford V8 Special, but an American 1936 Lincoln Zephyr.

Above: The Russian GAZ-AA ranked among the simplest and most widespread vehicles built under license from Ford and was always welcome as a captured vehicle. Here the engine is removed under field conditions in the midst of the Don plains by members of Artillery Regiment 171 in September of 1942. The "Lucky 71st Infantry Division" was wiped out at Stalingrad several months later.

The II./Mountain Artillery Regiment 79 won an unusual prize near Podwiskoye on August 6, 1941: a Russian GAZ 415 (Pikap), which was built from 1939 to 1941 (4 cylinders, 3282 cc, 50 HP) and based on the American Ford. The vehicle was repaired and used as a radio truck by the unit. Oberleutnant Trump stands at the shear telescope on October 4, 1941.

Schiffer Military History

Specializing in the German Military of World War II

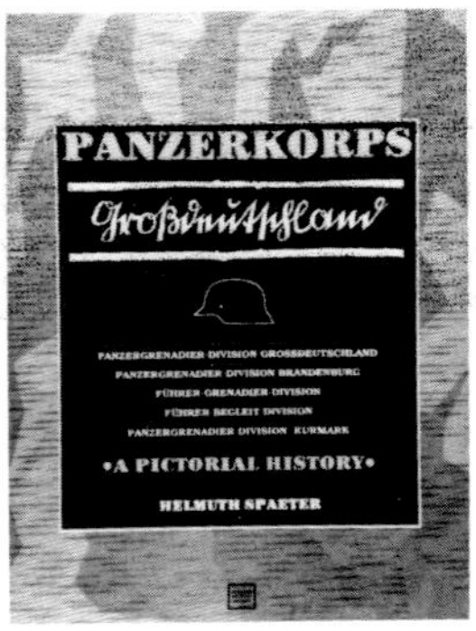

Panzerkorps *Grossdeutschland* Helmuth Spaeter. The elite *GD* Panzerkorps and its divisions, *GD, Brandenburg, Führer Begleit,* and *Kurmark* are covered in this volume, including their formation and battle histories during World War II. An extensive listing of *Knight's Cross* winners rounds out this comprehensive pictorial.
Size: 7 3/4" x 10 1/2" 248 pp.
Over 500 photos, maps, documents, charts
ISBN: 0-88740-245-3 hard cover $24.95

Stuka Pilot Hans-Ulrich Rudel Günther Just. Hans-Ulrich Rudel was the only soldier in World War II to receive the highest German decoration for bravery–the ***Golden Oak Leaves with Swords and Diamonds to the Knight's Cross***. This book by one of his fellow officers covers his entire life's story in many never before published photographs and reminiscences by family members, friends, and those who flew with him.
Size: 8 1/2" x 11" Over 400 photos 280 pp.
ISBN: 0-88740-252-6 hard cover $29.95

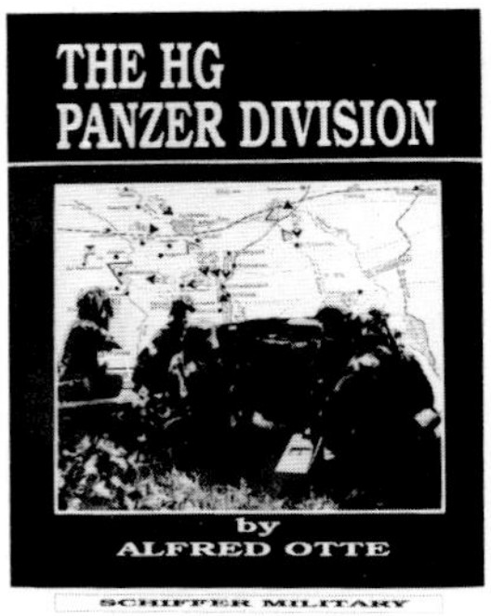

The *HG* Panzer Division Alfred Otte. This panzer division was formed from the Luftwaffe and took the name of its commander-in-chief, Herman Göring. It was originally formed as a motorized police unit, however, it later became a panzer regiment and eventually expanded to divisional size. The *HG* Panzer Division fought in the African Campaign, the Italian Campaign, on the eastern and western fronts and in the final defense of Germany. This is their story.
Size: 7 3/4" x 10 1/2" 176 pp.
325 b/w photographs, maps, documents
ISBN: 0-88740-206-2 hard cover $24.95

German Military Rifles and Machine Pistols 1871-1945 Hans Dieter Götz. This richly illustrated volume portrays the development of the modern German weapons and their ammunition, and includes many rare and experimental types. Among the weapons covered in this book are the Werder rifle, Mauser rifles, the various M/71 rifles and ammunition, the 88 cartridge, the Infantry Rifle 88, the 98 rifles, the Fallschirmjäger rifle, the 41 & 43 rifles, ERMA & Walther machine pistols and many more. Value guide included.
Size: 8 1/2" x 11" 248pp.
Many documents, drawings, and over 200 photos
ISBN: 0-88740-264-X hard cover $35.00

The Waffen-SS Herbert Walther. This interesting pictorial and factual documentation presents the history of the Waffen-SS from their initial formation to the end of the war. Over 500 photographs, many taken in the heat of battle, show the Waffen-SS divisions on many fronts, in victory and in defeat. This book explains in words and pictures the development of the Waffen-SS and their place in modern history.
Size: 7 3/4" x 10 1/2" 240 pp.
500+ b/w photographs
ISBN: 0-88740-204-6 hard cover $24.95